ZENBAKIEN ISTORIOA

THE NUMBER STORY

SMALL BOOK ONE

ENGLISH - BASQUE

---

*Numbers Teach Children
Their Number Names*

written and illustrated by

MISS ANNA

Early Reader Edition of *The Number Story 1*
Bronze Medal Winner, 2016 Wishing Shelf Book Award

Library of Congress Control Number: 2018902040

Names: Miss Anna, author.
Title: Number story : numbers teach children their number names / Miss Anna.
Description: Portland, OR: Lumpy Publishing, 2018.
Identifiers: ISBN 978-1-945977-94-7 | LCCN 2018902040
Summary: The pictures and rhymes present stories which introduce numbers 0-10.
Subjects: LCSH Numeration—English--Basque--Pictorial works--Juvenile literature. | BISAC JUVENILE NONFICTION /
Languages: English--Basque
Classification: LCC QA141.3 .M57 2018 | DDC 513—dc23

Publisher: Lumpy Publishing
Website: www.missannabooks.com
Email: missanna@missannabooks.com

Paperback: ISBN 978-1-945977-94-7
Printed in the U.S.A.   1 3 5 7 9 10 8 6 4 2

Zenbakien izenak ikasi
nahi al dituzue?

It is very easy and a lot of fun!

Oso erraza eta dibertigarria da!

Say-along our little jingle

Kanta dezagun elkarrekin istorio polit hau!

starting from Number One!

Has gaitezen Bat Zenbakiarekin!

1

ONE looks like my one finger.

BAT

nire bat-hatzaren antza du.

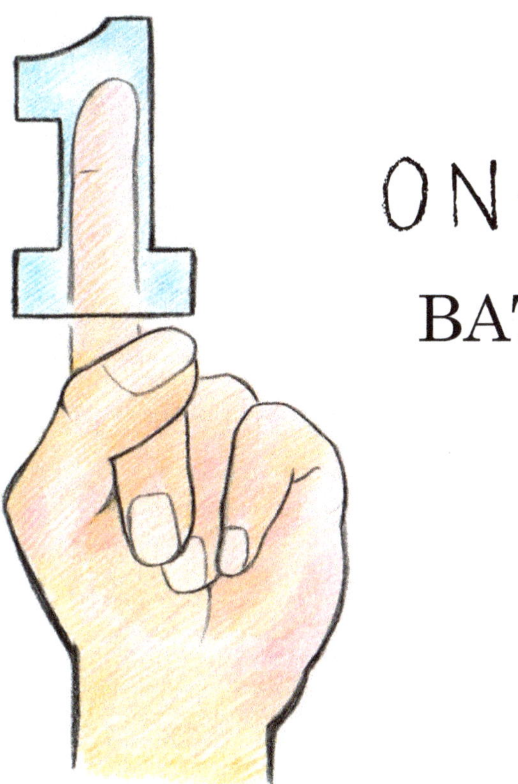

ONE!
BAT!

2

TWO   trails a tail.

BI

buztana duena da.

A TAIL! BUZTANA!

HIRU

koskorrak dituena da.

KOSKORRAK!

4

FOUR carries a sail.

LAU

itsasontzi-oihala duena da.

ITSASONTZI-OIHALA!

5

FIVE   is a racing track.

BOST

lasterketa bidea da.

VROOM
BRUUM!

SEI

barraskiloa bezala mugitzen da.

A SNAIL! BARRASKILOA!

7

SEVEN has a sharp angle.

ZAZPI

angelu zorrotza duena da.

BE CAREFUL! IT'S SHARP!
KONTUZ! ZORROTZA DA!

8

EIGHT is rollercoaster rails.

ZORTZI

errusiar mendi baten trenbidea da.

YUPII!
YIPPEE!

NINE is a bubble on a stick.

BEDERATZI

makil baten gainean

dagoen burbuila da.

A BUBBLE!

BURBUILA!

TEN is an eye of a whale.

HAMAR

bale baten begi bat da.

WINK!
KLISK!
HELLO! KAIXO!

And
Eta

0

ZERO   is an empty pail.

ZERO

ontzi hutsa da.

IT'S
EMPTY!
HUTSA DA!

Thank you for playing with us today.

We had a lot of fun too!

Eskerrik asko gurekin jolasteagatik.

Oso ondo pasa dugu guk ere!

We are your Number friends,
Zero to Ten,
Who will be here for you~
Zure Zenbaki lagunak gara
Zerotik Hamarrera.
Beti egongo gara hemen zuretzat.

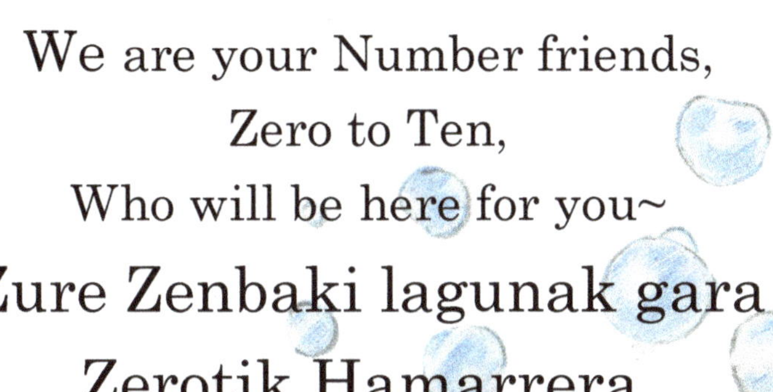

Bye-bye now!
See you again soon!
Agur Ben Hur!
Laster arte!

The Numbers are *SINGING* too!

To sing-a-long, look for Miss Anna Number Story
at your favorite music store like iTUNES.

MP3

Numbers 0-10
IDENTIFYING
& COUNTING

Numbers 11-20
& Ordinals
first, second, third...

Numbers 0-100
& Place Values
ones, tens, hundreds...

About Clocks
& Telling Time
hours, minutes, seconds

Number Story 1 & 2
isbn: 978-0-996216-48-7

Number Story 3 & 4
isbn: 978-1-945977-01-5

Number Story 5 & 6
isbn: 978-1-945977-06-0

Number Story 7 & 8
isbn: 978-1-949320-40-4

For more Miss Anna books to love,
visit us at

www.missannabooks.com

Numbers are working hard all over the world!
Come Travel the World with Us!